Fire and Explosions at Rocket Fuel Plant
Henderson, Nevada

Investigated by: J. Gordon Routley

This is Report 021 of the Major Fires Investigation Project conducted by TriData Corporation under contract EMW-8-4321 to the United States Fire Administration, Federal Emergency Management Agency.

Homeland Security

Department of Homeland Security
United States Fire Administration
National Fire Data Center

U.S. Fire Administration Fire Investigations Program

The U.S. Fire Administration develops reports on selected major fires throughout the country. The fires usually involve multiple deaths or a large loss of property. But the primary criterion for deciding to do a report is whether it will result in significant "lessons learned." In some cases these lessons bring to light new knowledge about fire--the effect of building construction or contents, human behavior in fire, etc. In other cases, the lessons are not new but are serious enough to highlight once again, with yet another fire tragedy report. In some cases, special reports are developed to discuss events, drills, or new technologies which are of interest to the fire service.

The reports are sent to fire magazines and are distributed at National and Regional fire meetings. The International Association of Fire Chiefs assists the USFA in disseminating the findings throughout the fire service. On a continuing basis the reports are available on request from the USFA; announcements of their availability are published widely in fire journals and newsletters.

This body of work provides detailed information on the nature of the fire problem for policymakers who must decide on allocations of resources between fire and other pressing problems, and within the fire service to improve codes and code enforcement, training, public fire education, building technology, and other related areas.

The Fire Administration, which has no regulatory authority, sends an experienced fire investigator into a community after a major incident only after having conferred with the local fire authorities to insure that the assistance and presence of the USFA would be supportive and would in no way interfere with any review of the incident they are themselves conducting. The intent is not to arrive during the event or even immediately after, but rather after the dust settles, so that a complete and objective review of all the important aspects of the incident can be made. Local authorities review the USFA's report while it is in draft. The USFA investigator or team is available to local authorities should they wish to request technical assistance for their own investigation.

For additional copies of this report write to the U.S. Fire Administration, 16825 South Seton Avenue, Emmitsburg, Maryland 21727. The report is available on the Administration's Web site at http://www.usfa.dhs.gov/

U.S. Fire Administration

Mission Statement

As an entity of the Department of Homeland Security, the mission of the USFA is to reduce life and economic losses due to fire and related emergencies, through leadership, advocacy, coordination, and support. We serve the Nation independently, in coordination with other Federal agencies, and in partnership with fire protection and emergency service communities. With a commitment to excellence, we provide public education, training, technology, and data initiatives.

TABLE OF CONTENTS

FIRE AND EXPLOSIONS AT ROCKET FUEL PLANT
Henderson, Nevada
May 4, 1988

Local Contacts: Chief Roy Parrish
Deputy Chief John Pappageorge
Captain Robert James, Fire Investigations Unit
Clark County Fire Department
707 East Desert Inn Road
Las Vegas, Nevada 89109
(702) 455-7311

OVERVIEW

A series of explosions on May 4, 1988, near the city of Henderson, Nevada, claimed two lives, injured approximately 327 people, including 15 firefighters, and caused damage estimated over 100 million dollars. The explosions affected a large portion of the metropolitan Las Vegas area and caused the activation of disaster plans by several agencies. Considering the magnitude of the explosions, the loss of only two lives, and the fact that only a few of the injuries were critical, can be described as very fortunate. The incident presented tremendous risk and unusual challenge to the fire departments involved, but they managed the incident with relatively minor casualties. The lives of most of the plant employees were saved by their decision to evacuate the plant, prior to the major explosions. This and other key issues are summarized in the table on the following page.

BACKGROUND

The incident occurred in an unincorporated industrial area of Clark County, approximately 10 miles southeast of downtown Las Vegas. The specific location is a county "island" surrounded by the city of Henderson, a rapidly growing suburb with a population of 50,000. The Henderson area has been a center of industrial production, much of it related to defense industries, since World War II.

SUMMARY OF KEY ISSUES

Issues	Comments
Ignition	Welding torch ignited structure of chemical plant.
Construction	Fiberglass walls in steel-frame building helped spread fire.
Hazardous Materials (Hazmat)	Plant produced and stored ammonium perchlorate, an oxidizer in rocket fuels. Other Hazmat, including anhydrous ammonia, hydrochloric acid, and nitric acid, were present in bulk quantities.
Interagency Coordination	Working relationships established among fire, police, and other agencies as a result of previous hotel fires helped make management of this incident more efficient. An on-scene command post and a multi-jurisdictional command center worked well.
Public Information	News media demand for immediate information overwhelmed official sources; result was misinformation in the media, widespread rumors, and near panic.
Industrial Safety Planning	Employees saved their lives by rapid evacuation once the danger of explosions was recognized. Lack of an evacuation plan made it difficult to account for all employees after the incident.
Land Development	Industrial development close to the plant was destroyed and damage was reported over a large area. Residential development near the plant was damaged, but few injuries occurred there because of the time of day.
Emergency Medical Triage	Triage points set up outside hospitals kept them from being overwhelmed as the widespread injured came in for aid.
Hazmat Guidebook	Department of Transportation Emergency Response Guidebook information on large particle size ammonium perchlorate may imply less danger than exists, in light of this incident.

The Pacific Engineering Production Company of Nevada (PEPCON), site of the explosions, is one of only two free world producers of ammonium perchlorate, an oxidizer used in solid fuel rocket boosters, including the Space Shuttle and military weapons. The other producer also is located within Clark County, less than 1.5 miles away from the PEPCON facility – within the area that suffered some blast damage.

The plant occupied approximately eight acres, including six buildings and outside chemical storage and process areas. The plant was constructed in the 1950's in an isolated desert area. The isolation was reduced by rapid growth in the metropolitan Las Vegas area over the last decade.

At the time of the incident, there was a large marshmallow factory within 500 feet of the plant, and a gravel quarry was in operation nearby. While the closest residential and other commercial occupancies were approximately one and one half to two miles away, urban growth had greatly increased the population in the area affected by the explosions.

HAZMAT

Ammonium perchlorate was the only product manufactured at the PEPCON facility. The process uses several Hazmat, including anhydrous ammonia, hydrochloric acid, nitric acid, and various chlorate compounds. These chemicals were shipped to the site primarily by rail and were present in bulk quantities. An estimated 8.5 million pounds of the finished product were stored at the facility.

Ammonium perchlorate is a powerful oxidizer which is mixed with combustible materials to produce rocket fuels. The mixed fuels provide a high energy release at a very rapid rate of combustion.

The rate of combustion is controlled by the mixture, particle size, and moisture content. Finely ground aluminum frequently is used as the combustible component.

Without the combustible component, ammonium perchlorate is classified as an oxidizer and considered less hazardous than a mixed fuel. Normal combustion will be greatly accelerated in its presence, and contamination of ammonium perchlorate with an organic product will create explosive mixtures.

It is normally shipped in large aluminum "tote" containers, each holding several thousand pounds of the white granular material. Several hundred of the aluminum "totes" were stored in one area of the plant awaiting shipment, along with a smaller number of fiber drums. A quantity of nearly empty drums was also on hand to be refilled.

Another storage area at the plant contained several thousand 55-gallon plastic drums of the product waiting final blending to customer specifications. The plastic drums were not used to ship the product outside the plant.

The product (ammonium perchlorate) apparently had not been tested for mass (large quantity) detonation prior to this fire, and its classification was based on small scale tests. Although not previously considered to be explosive, this incident obviously gives testimony to the fact that ammonium perchlorate can explode.

According to the U.S. DOT *Emergency Response Guidebook* (the "yellow book" carried by many fire companies), the precautions for ammonium perchlorate are somewhat different depending on its particle sizes. There are different guide numbers, Hazmat numbers, and instructions for particle size under 45 microns and over 45 microns. A firefighter approaching a vehicle or plant might well not know which instructions to follow without detailed knowledge gained in pre-fire planning. At the Henderson plant, particle sizes were 90 microns and above, according to Deputy Chief Pappageorge of the Clark County Fire Department.

The information in the guidebook is intended to apply to quantities that would be encountered in transportation. For both particle sizes, the guide clearly notes the possibility of fire or explosion. For the smaller particle size, it provides the same information as a Class A explosive. It advises to stop traffic and evacuate to one mile away and not to fight the fire.

For the larger particle size product, the guidebook suggests the use of master streams from a safe distance and warns of irritating or poisonous gases that may be produced from a fire.

For either particle size, the firefighters are warned of the explosion potential. Bulk quantities as large as those normally stored in a plant or fixed site are not covered by the manual, but the warnings are dire enough to preclude approaching a major fire involving the product. However, the guidebook may not make it clear enough that even large particle size ammonium perchlorate can explode catastrophically.

In addition to the chemicals at the plant, a 16-inch, high-pressure (300 psi) natural gas transmission line ran underneath the plant and also supplied the plant through a pressure reducing assembly.

THE FIRE

The fire is reported to have originated in or around a drying process structure in the PEPCON plant between 1130 and 1140. The steel frame with fiberglass walls and roof structure had been damaged in a windstorm and employees were conducting repairs using a welding torch at the time. The fire spread rapidly in the fiberglass material, accelerated by ammonium perchlorate residue in the area.

As employees attempted to fight the fire with hoselines, the flames spread to 55-gallon plastic drums containing the product that was stored next to the building.

The employee efforts at extinguishment were unsuccessful, and they abandoned the effort when the first of a serious of explosions occurred in the 55-gallon drums. The time between ignition and the first explosion has not been determined exactly; it was estimated at 10-20 minutes. When the control efforts were abandoned, most of the plant employees evacuated the area by running or driving away. Approximately 75 managed to evacuate, leaving only the two who were killed in subsequent larger explosions. One of these victims stayed behind to call the Clark County Fire Department and the other was confined to a wheelchair and was unable to leave the area. The first explosion also alerted employees of the nearby marshmallow factory, and they also evacuated the area.

The fire continued to spread in the stacks of filled 55-gallon plastic drums and created an extremely intense fireball. The first of two major explosions then occurred in the drum storage area. The fire continued to spread and reached the storage area for the filled aluminum shipping containers. This resulted in an even larger, second major explosion, approximately four minutes later. Very little fuel remained after the second explosion and the flame diminished rapidly except for the flame plume created when the high pressure natural gasline beneath the plant was ruptured in one of the explosions. The gasline was shut off at 1259 hours by the gas company, at a valve about a mile away, eliminating the fuel for this fire.

A huge column of smoke rose from the plant and was carried downwind to the east, over most of the residential and business areas of Henderson. The smoke rose on the thermal column to an altitude of several thousand feet and was spotted almost 100 miles away.

All told, seven explosions occurred involving various containers of ammonium perchlorate, with the two largest occurring in the plastic drums and then the aluminum containers. These two explosions were measured at 3.0 and 3.5 on the Richter scale at an observatory in California! Over eight million pounds of the product were consumed in the fire and explosions. A crater estimated at 15 feet deep and over 200 feet long was left in the storage area.

FIRE DEPARTMENT RESPONSE

The Clark County Fire Department had received numerous telephone calls reporting the fire after the first (small) explosion, including the call from the plant employee. These calls had begun at 1151 and a first alarm assignment was dispatched immediately. The closest Clark County units had a response of over five miles and could see the heavy smoke from their station.

At approximately the same time, the huge column was spotted by the fire chief of the city of Henderson who was leaving the main fire station, approximately 1.5 miles north of the PEPCON facility. The chief immediately ordered his units to be dispatched and headed toward the scene. As he approached within a mile of the plant, he could see a massive white and orange fireball, approximately 100 feet in diameter, and dozens of people running across the desert toward him. He advised his dispatcher to call for mutual aid assistance although he was still unaware of the nature of the fire.

As he approached the scene at 1154, the first of the two major explosions occurred. The shock wave shattered the windows of his car and showered the chief and his passenger with glass. The driver of a heavily damaged vehicle coming away from the plant advised the chief of the danger of further, even larger explosions. With this information, the chief turned around and headed back toward his station. The other Henderson companies en route to the scene stopped where they were on their own volition after the explosion (about one mile away).

Approximately four minutes after the first major explosion, the second large explosion occurred. Witnesses reported that this explosion created a visible shock wave coming toward them across the ground. Several videotape recordings of the explosions were made by people in the area, graphically demonstrating the movement of the shock wave.

The second major explosion virtually destroyed the chief's car. The chief and his passenger were cut by flying glass, but he was able to drive the damaged vehicle to a hospital to seek treatment. The windshields of the responding Henderson Fire Department apparatus were blown in, and the drivers and officers were injured by the shattered glass. The Henderson Fire Department was essentially totally incapacitated by the second major explosion. The injuries consisted of numerous cuts from flying glass, but did not require hospitalization.

The Clark County response was upgraded to a third alarm while units were still en route. Several area fire departments also responded on mutual aid. The Clark County units staged 1.5 miles from the scene and provided assistance to the injured Henderson firefighters. From this distance they attempted to size-up the situation. Both the PEPCON facility and the neighboring marshmallow plant had been destroyed in the explosions prior to their arrival. The magnitude of the fire in the PEPCON facility was beyond any fire suppression capability, and flames also were visible in the rubble of the marshmallow plant. The only hydrants were in the immediate area of the two involved plants, but there was no water supply due to the loss of electrical power to the pumps. Recognizing the danger and futility of operations, no attempt was made to approach or to fight the fire.

A command post was established more than two miles from the scene at a location that afforded a view of the involved area. Responding fire and medical units were staged as an assessment was made of the situation.

FIRE DEPARTMENT OPERATIONS

The immediate concerns for emergency response personnel were:

1. The danger of additional explosions.

2. The possibility of toxic products being released with the smoke.

3. The need to search for and treat victims in the immediate area.

4. The need for damage assessment and emergency medical treatment in the entire area affected by the explosions.

A decision was made to evacuate a 5-mile radius around the plant, concentrating on the downwind direction as the priority. This assignment was given to the Las Vegas Metropolitan Police Department, assisted by the Nevada State Police, and, later, the National Guard. The roads in the area were clogged in both directions with residents trying to leave and curious spectators headed toward the scene. The massive traffic jams took over two hours to clear.

The command post established by the Clark County Fire Department was the focal point of operations in the immediate area. The department's Hazmat Response Team attempted to make an assessment of the toxicity danger using air sampling instruments. Assistance in air sampling and evaluation also was provided by Nellis Air Force Base Fire Department personnel and locally based experts from the military, U.S. Department of Energy, and both the State and Federal Environmental Protection Agency (EPA). The plant manager was located and brought to the command post to assist in evaluating the situation.

A conclusion was reached more than an hour after the first explosions that the airborne products would be a respiratory irritant, but not highly toxic, and that the danger of further explosions was remote. Expansion of the evacuation zone to 10 miles was being considered, but implementation was cancelled on the basis of this information, though several cases of respiratory irritation were reported in a small community approximately 30 miles downwind.

After the fires began to subside, a battalion chief and the fire inspector who was familiar with the plant made an initial, close-in survey and determined that there was no further risk of explosions. Overhaul was extremely difficult, since water had to be trucked in and constant evaluation of the dangers from Hazmat was necessary.

Crews donned protective clothing near the command post and were transported into the scene using self-contained breathing apparatus (SCBA). Leaking tanks of anhydrous ammonia and residue from acids and other products made progress slow and required continuing evaluation by the Clark County Fire Department Hazmat Team. During the overhaul stages, several firefighters required treatment for respiratory irritation. Overhaul continued until dusk and was resumed the following day.

During the overhaul process the remains of one plant employee were located. No trace of the second victim was ever found.

EMERGENCY MEDICAL OPERATIONS

Emergency medical treatment in the area of the explosions consisted mainly of basic life support for those injured by flying debris. The damaged cars in the area were searched and one critical (head trauma) patient was located in a car 1/2 mile from the PEPCON facility. Employees, who had run in all directions, were assembled in an area adjacent to the command post. Over 100 employees were on the premises of the PEPCON and marshmallow plants when the incident began, but only 20 to 30 required hospital treatments, and most were released within two hours. It took over six hours to account for all of the employees and determine that two were missing.

Both the fire departments and a large private-sector ambulance service provide emergency medical treatment and transportation in the metropolitan Las Vegas area and work together effectively on a routine basis. The ambulance service established a triage sector to manage patients at the scene, located near the fire department command post. The triage sector was staffed by paramedics from both private and public units and dealt mainly with the employees and others injured in the immediate area.

The major emergency medical problem was the estimated 300 patients who were injured in the surrounding area. These patients were distributed over an area of 50 to 75 square miles and suffered injuries primarily from flying glass and falling debris from ceilings and light fixtures. One infant was seriously cut by glass from a broken window, more than two miles from the scene.

Emergency medical services (EMS) treated and transported approximately 100 patients to five hospitals in the region. The remaining 200 to 300 patients presented themselves to hospitals as "walking wounded." Triage areas were set up outside hospitals to handle this influx in an orderly manner. The hospitals had activated their disaster plans, making trauma teams and operating rooms ready, but they received only a handful of seriously injured patients. Emergency room facilities were taxed by the numbers but not the severity of the injuries.

The closest hospital, St. Rose de Lima in downtown Henderson, treated over 100 patients, many in the parking lot. The hospital had many windows broken by the explosion and was operating on emergency generator power. The hospital continued to provide treatment while making preparations to evacuate, in case this became necessary.

Approximately four hours after the incident began, the hospitals were advised by the fire department that their disaster plans could be deactivated.

A total of 15 firefighters were injured, eight from flying broken glass and seven from respiratory difficulties during overhaul.

DAMAGE ASSESSMENT

Both the PEPCON and marshmallow manufacturing facilities were virtually destroyed. Damage with a 1.5 mile radius was heavy, including destroyed cars, structural damage to buildings and downed power lines. Within three miles there was extensive window breakage and moderate structural damage. Many structures had damage to suspended ceilings and overhangs, windows and doors, exterior details, and cracked walls.

Damage extended for a radius of up to 10 miles. Buildings were damaged throughout Henderson including over 100,000 dollars damage to the main fire station and heavy structural damage to a warehouse next door. Hundreds of windows were shattered, doors were blown off their hinges, walls cracked, and scores of people were injured by flying glass and debris. At Las Vegas' McCarran International Airport seven miles away, windows were cracked and doors were pushed open. A Boeing 737 on final approach was buffeted by the shock wave.

The fire departments in the area were heavily committed to the actual incident scene and had little involvement with damage assessment or other activities away from the immediate area.

INFORMATION MANAGEMENT

One of the major challenges faced by the Clark County Fire Department in this incident was the management of information. The department itself had an urgent need for information on what had happened, was happening, and could happen, in order to formulate a plan for operations and evacuation. This required consultation with fire department personnel, plant management, and experts from other agencies, under extremes of stress and uncertainty.

While the process of planning and evaluation was taking place, there were immediate and constant pressures from the local news media for details and for information to broadcast to the public concerning the dangers and actions that should be taken. The time required to gather and analyze information resulted in some incorrect information being broadcast and caused widespread public confusion. At the same time the national news media were calling for more details. The Clark County Fire Department's public information officer responded and established an official source of media information within an hour after the explosion.

Many residents were in near panic from rumors of several different scenarios and dangers. Radio and television stations quickly devoted their air time to the situation, but lacked a source of accurate information during the first hour. Conflicting information was broadcast and, as a result, people in the area reported confusion about whether to stay indoors to avoid the smoke, evacuate, go to shelters, or take some other action. The confusion extended to schools in the area, with some keeping children inside and others sending students home.

Telephone lines were overloaded with people checking on each other's welfare, seeking advice from different sources, or trying to report conditions to emergency response agencies and the news media. The 9-1-1 telephone system was rapidly overloaded with concerned callers, many seeking instructions, and the cellular telephone system was overloaded.

This emphasizes the need to establish working lines of communication with the news media. But even with a good relationship, the ability to provide accurate information and to depend upon the news media to convey instructions to the public, becomes very uncertain in an incident of this magnitude. In this case several inaccurate reports, including "confirmed reports of 9 to 14 dead" were broadcast.

There was also a concern for the safety of news gathering personnel who approached closer to the involved area than fire department personnel would venture, including helicopters circling within the danger zone.

INTERAGENCY COOPERATION

The Clark County Fire Department received valuable assistance from a number of other agencies during this incident. This included mutual aid from several fire departments, including Henderson, Las Vegas, Boulder, Nellis Air Force Base, and the U.S. Park Service. The fire department response was effectively coordinated through the on-scene command post and the communications center that serves the city of Las Vegas, Clark County, and North Las Vegas. This routine approach to mutual aid and automatic assistance also facilitated the cooperation between the public and private EMS providers. Effective planning also provided for the coordinated activation of hospital disaster plans.

The fire department received additional assistance, including extra SCBAs loaned by hotels and casinos for use by off-shift firefighters. Lighting plants were provided by Nellis Air Force Base, and a private catering firm fed all personnel at the scene, according to an established plan.

Hazmat expertise was provided by several agencies, primarily military and other Federal government agencies that are active in the area.

With fire department capabilities concentrating on the fire and Hazmat situations, the police department managed evacuation and control of traffic and spectators. The Governor of the State of Nevada responded to the scene and activated the National Guard to assist in securing the evacuated area. All of these functions were effectively coordinated through liaisons at the command post.

Several agencies, including the Red Cross and school districts, were involved in providing temporary shelter for evacuees.

LESSONS LEARNED

1. **Land development decisions must consider risks of disasters.**

   The potential destructive power from an incident of this type needs to be evaluated in land use decisions. The encroachment of residential and commercial development into the area around the PEPCON plant contributed significantly to the injuries and damage. The magnitude of the incident was much greater than had been contemplated by urban planners or in pre-incident planning.

2. **Need for triage outside hospitals: large numbers of even minor injuries can overwhelm a medical facility once inside.**

Damage and injuries spread over a large area present unusual challenges to emergency services, which are accustomed to incidents occurring in a well-defined area. Large numbers of injured presented themselves to hospitals. Triage centers need to be set up outside hospitals to prevent overloads within the hospitals when the EMS cannot "capture" most victims at the site of the incident. Fortunately in this incident there were not large numbers of seriously injured waiting for assistance and relying on public agencies for treatment and transportation. This potential needs to be considered more than it has been by local communities.

3. **Disaster mutual aid plans should be established, practiced, and kept up-to-date.**

The value of established mutual aid and interagency coordination procedures was demonstrated once again. Many of these relationships came as a result of Las Vegas' experience from major fires at the MGM Grand and Hilton Hotels in 1980 and 1981. Communities should review their disaster coordination plans and make sure they are up-to-date.

4. **Hazmat incidents require size-up from a safe distance.**

The need for a "stand back and assess the situation" strategy for some Hazmat incidents was well demonstrated. Had fire units continued at full speed to the scene they probably would have been destroyed.

5. **Public information needs to be accurate and timely in a disaster.**

In spite of a good relationship and established procedures, dealing effectively with the media was a major problem in the early stages of the incident. Misinformation by the media and rumors among the public created near panic. Since the aftermath was not dangerous, it did not matter much here whether people stayed indoors or not. But in an environmentally serious incident, clear information should be given to the public as soon as possible on what to do, even if that must be changed as conditions change. The departments did a good job in providing information to the media as it became available, but the media did not wait for good information.

6. **Evacuation plans and implementation must consider human nature and the media.**

The roads in the area were virtually gridlocked by spectators going toward the scene and local residents fleeing. Media misinformation, the failure to broadcast adequate, specific requests to stay away from the scene, and the inability to control the roads early enough, exacerbated the traffic situation. The need for traffic control around high hazards should be considered in disaster plans.

7. **The problem of assessing the immediate risk of Hazmat releases and products of combustion on the surrounding area needs additional research and development.**

Determining the risk of explosion and the risk of toxic fumes to the public and to their firefighters can be extremely difficult. Some aspects of risk assessment are too specialized to be covered in general Hazmat training courses. Special expertise needs to be called into play when unusual or exotic Hazmat is known to be present in a plant or other location. The "worst case" situation needs to be anticipated.

Local fire departments also need to know who to call for quick assistance in air sampling. Whatever the designated agency – most often it is the State EPA – its personnel need to be equipped, trained, and prepared to respond quickly to locations throughout their State if they are to be of real assistance. (In the Nanticoke, Pennsylvania, chemical plant fire, which occurred in March of 1987 and was also investigated by the USFA, getting assistance in air sampling was a major problem.)[1]

Sampling devices to identify gases need to be improved and put into greater use. Improved methods need to be developed to make a more rapid assessment of risk.

8. Industrial safety plans need to be established, kept up-to-date, and understood by all employees.

The employees in the chemical plant averted a life loss catastrophe by fleeing immediately after the first ("small") explosion and fireball. There was no alarm system and no evacuation plan, according to the employees. Plants such as this should have explicit emergency plans, and all employees should be trained.

9. Safety of the handicapped needs to be considered in high-hazard occupancies.

One of the two fatalities in this fire was a wheelchair-bound employee who obviously could not just run across the desert or jump in his car, as most others did. Society now encourages and assists the handicapped to visit and work in a much wider range of occupancies, which exposes them to new risks. The handicapped need to be given realistic appraisal of their risks in a potential emergency and their alternatives for escape or refuge, even though the probability of the event occurring is low. Volunteers might be assigned to handicapped individuals to help them to escape by car or truck, or by being wheeled or carried to a safe distance. This may suffice for most emergencies, but in the face of a catastrophic explosion the value of such assignments may be moot.

10. The section on ammonium perchlorate in the DOT Emergency Response Guidebook needs to be revised.

The danger of explosion from large (over 45 microns) particle size ammonium perchlorate when exposed to flames seems to be much greater than indicated in the manual. The small particle size might be viewed as a Class A explosive and the large particle size version as almost that dangerous.

---

[1] See "Evacuation of Nanticoke, Pennsylvania, Due to Metal processing Plant Fire (March 24, 1987)", USFA, 1987.

LIST OF SLIDES/PHOTOS

Slides are included with the master report at the USFA. Copies made from enlarged photographs from the slides are on the following pages. They are keyed to the slide captions below. (The four asterisked items were somewhat redundant and/or not clear enough to warrant reproduction.)

1. Aerial photo of PEPCON – before explosion.

2. Area map (not to scale) – red line is gas transmission line.

3. Developing fire sequence – prior to first explosion – from Marshmallow Plant.

*4. Developing fire sequence – prior to first explosion – from Marshmallow Plant.

5. Developing fire sequence – prior to first explosion – from Marshmallow Plant.

6. Immediately after first explosion (note insulation truck in Marshmallow Plant lot undamaged at this time).

7. One of the two large explosions.

8. Same explosion a moment later (note dust from shock wave).

9. Flame from ruptured natural gas line taken after second explosion -- estimated height 200 feet.

*10. Views of PEPCON Plant damage – white crystal/powder is residue of ammonium perchlorate.

*11. Views of PEPCON Plant damage – white crystal/powder is residue of ammonium perchlorate.

12. Views of PEPCON Plant damage – white crystal/powder is residue of ammonium perchlorate.

13. Views of PEPCON Plant damage – white crystal/powder is residue of ammonium perchlorate.

14. Anhydrous ammonia tank cars adjacent to PEPCON Plant.

15. Damage around Marshmallow Plant.

16. Damage around Marshmallow Plant.

17. Damage around Marshmallow Plant.

*18. Damage around Marshmallow Plant.

19. Damaged cars (employee parking lot).

20. Truck damaged by blasts (same truck as in Photo 6).

21. Railroad crossing – approximately 1/2 mile from PEPCON Plant.

22. Damage to warehouse roof – approximately 1 mile from PEPCON Plant.

23. Damage to warehouse roof – approximately 1 mile from PEPCON Plant.

1. Aerial photo of PEPCON – before explosion.

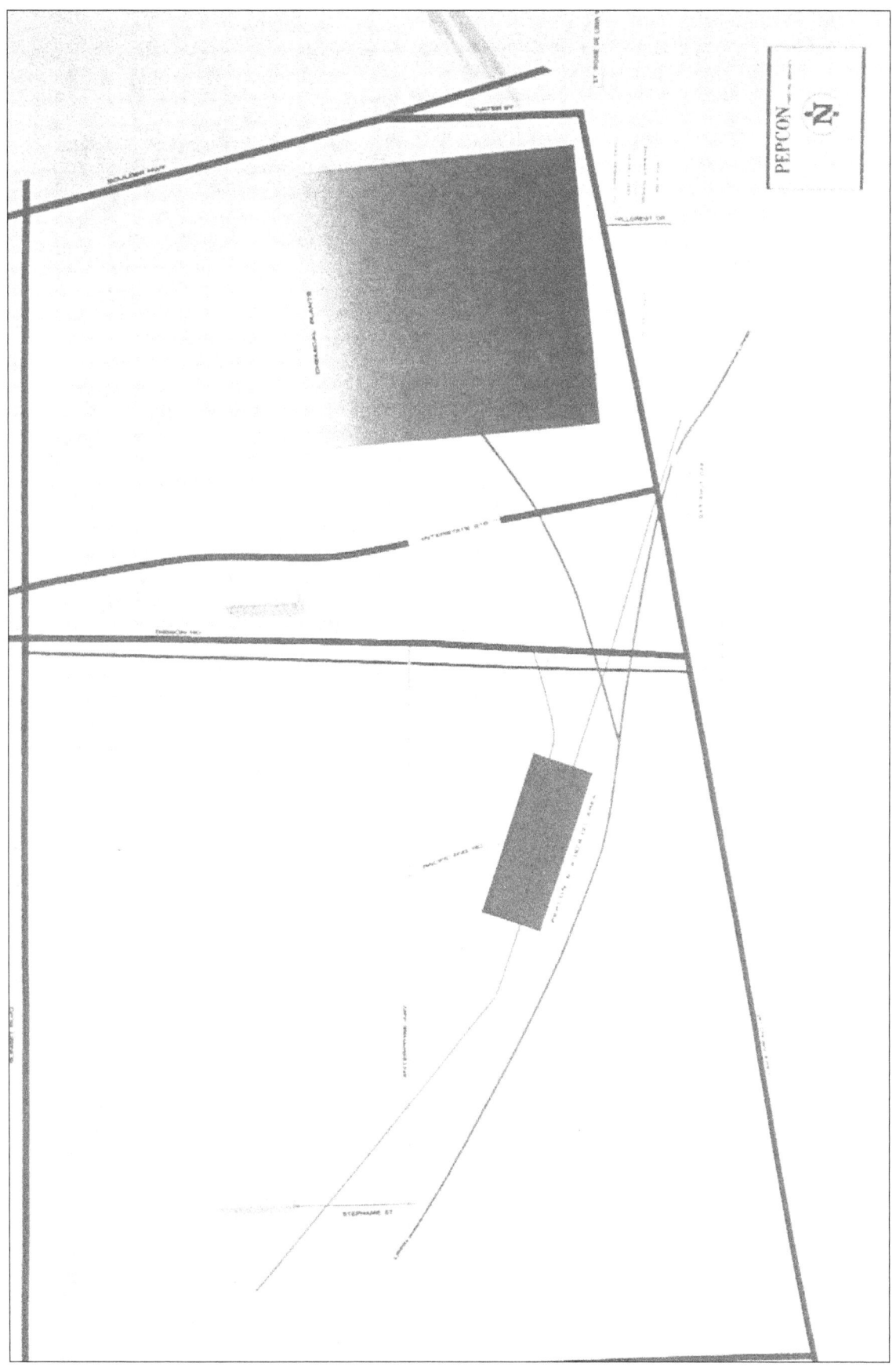

2. Area map (not to scale) – red line is gas transmission line.

3. Developing fire sequence – prior to first explosion – from Marshmallow Plant.

5. Developing fire sequence – prior to first explosion – from Marshmallow Plant.

6. Immediately after first explosion (note insulation truck in Marshmallow Plant lot undamaged at this time).

7. One of the two large explosions.

8. Same explosion a moment later (note dust from shock wave).

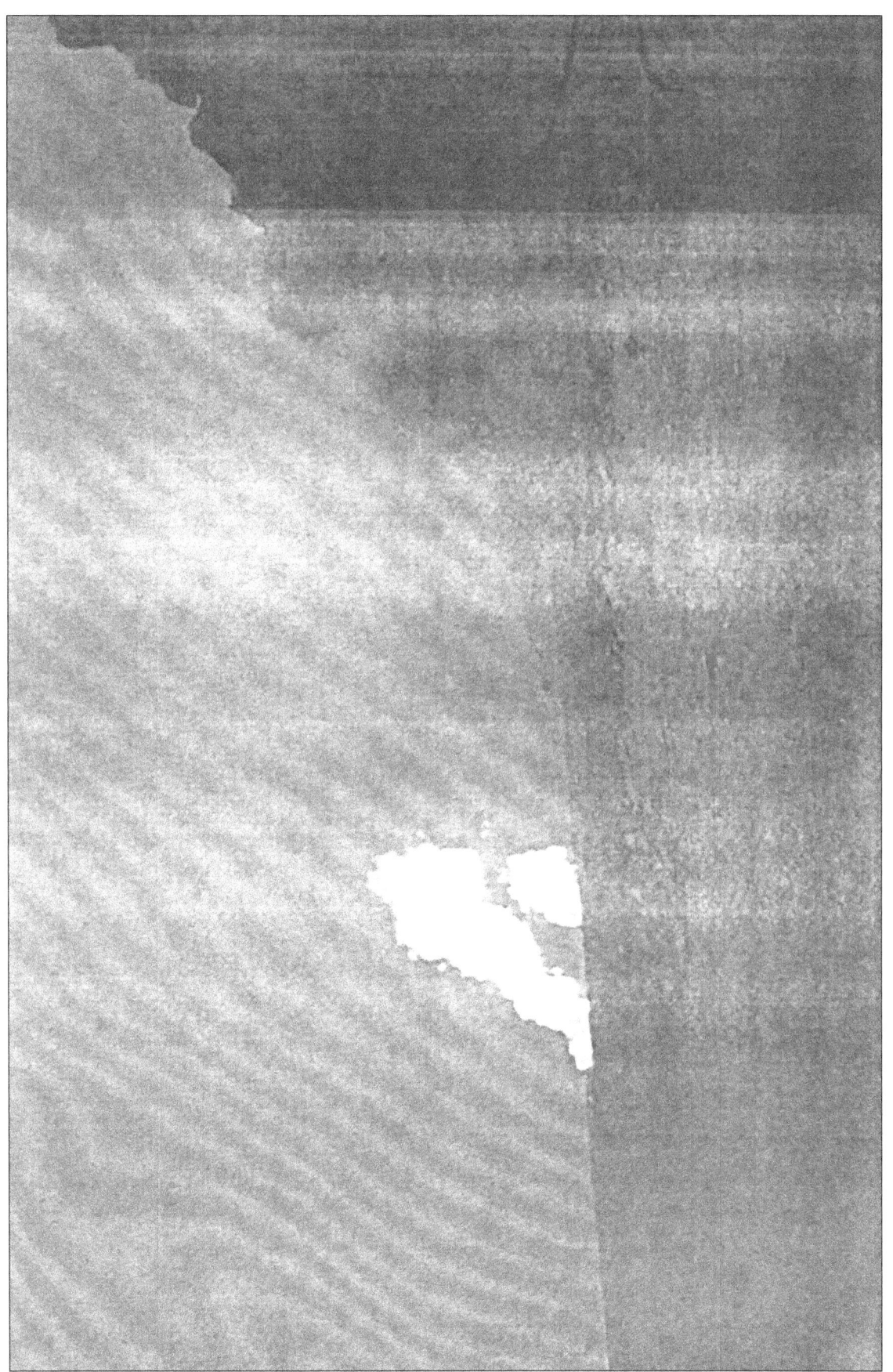

9. Flame from ruptured natural gas line taken after second explosion -- estimated height 200 feet.

12. Views of PEPCON Plant damage – white crystal/powder is residue of ammonium perchlorate.

13. Views of PEPCON Plant damage – white crystal/powder is residue of ammonium perchlorate.

14. Anhydrous ammonia tank cars adjacent to PEPCON Plant.

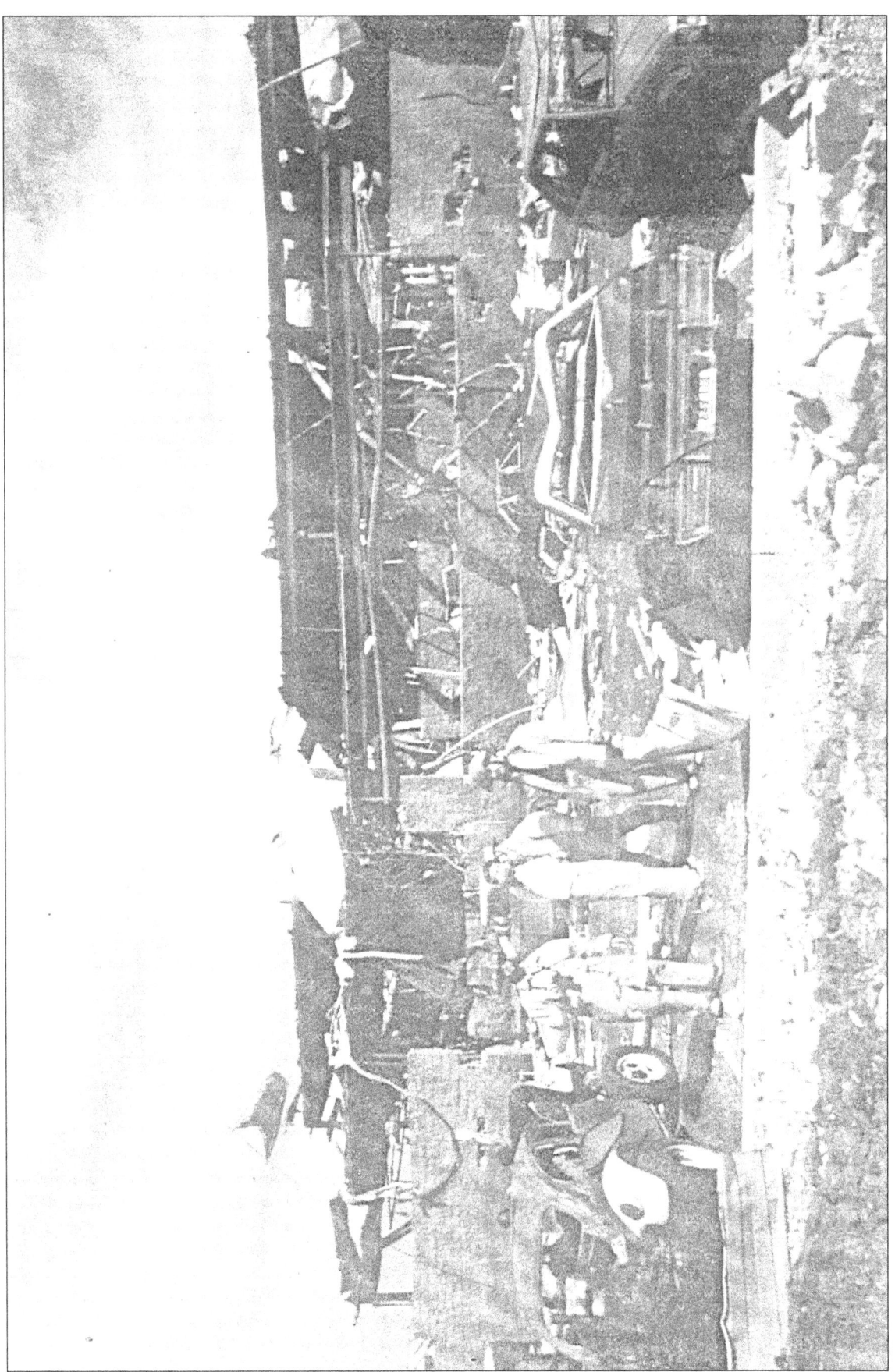

15. Damage around Marshmallow Plant.

16. Damage around Marshmallow Plant.

17. Damage around Marshmallow Plant.

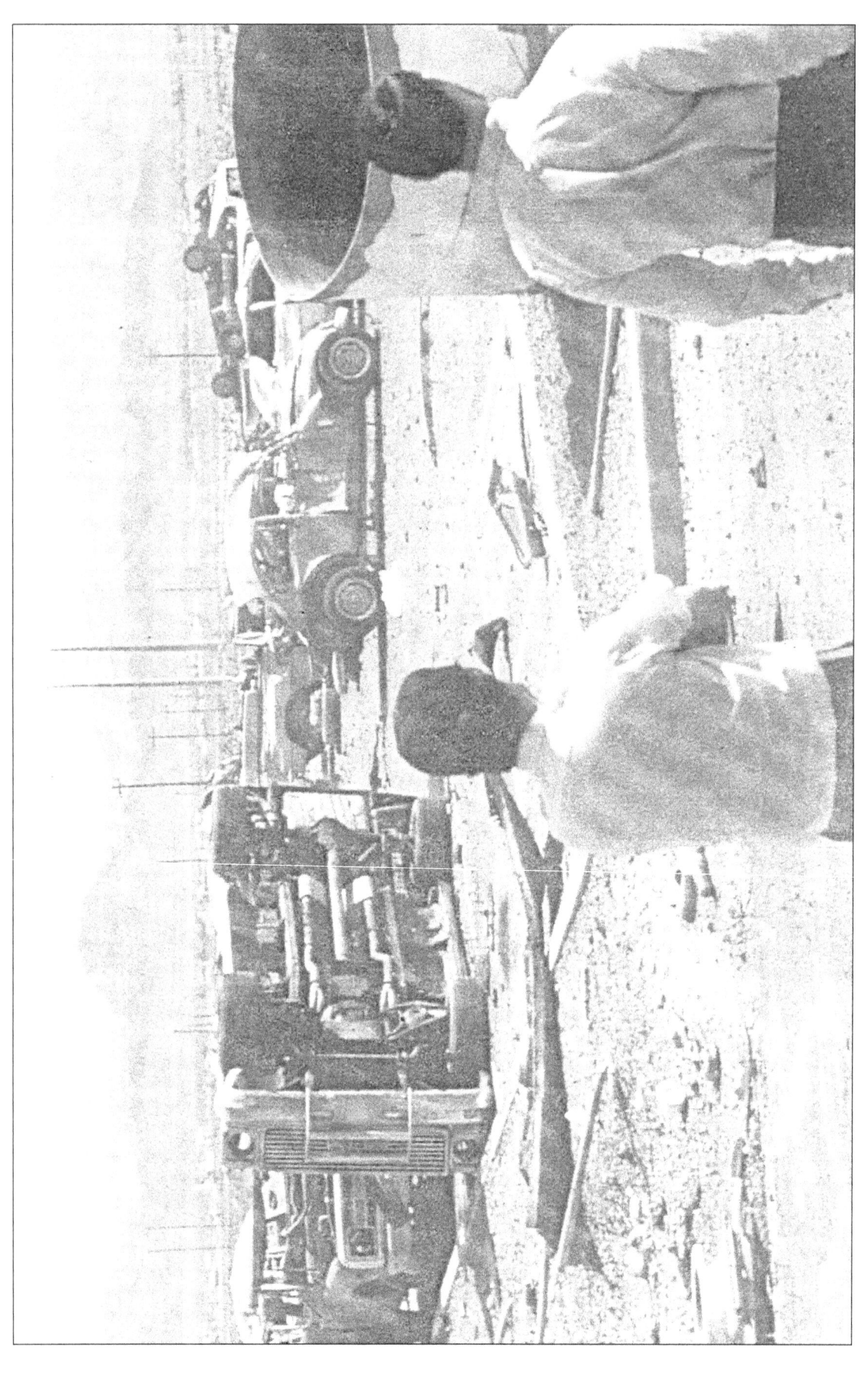

19. Damaged cars (employee parking lot).

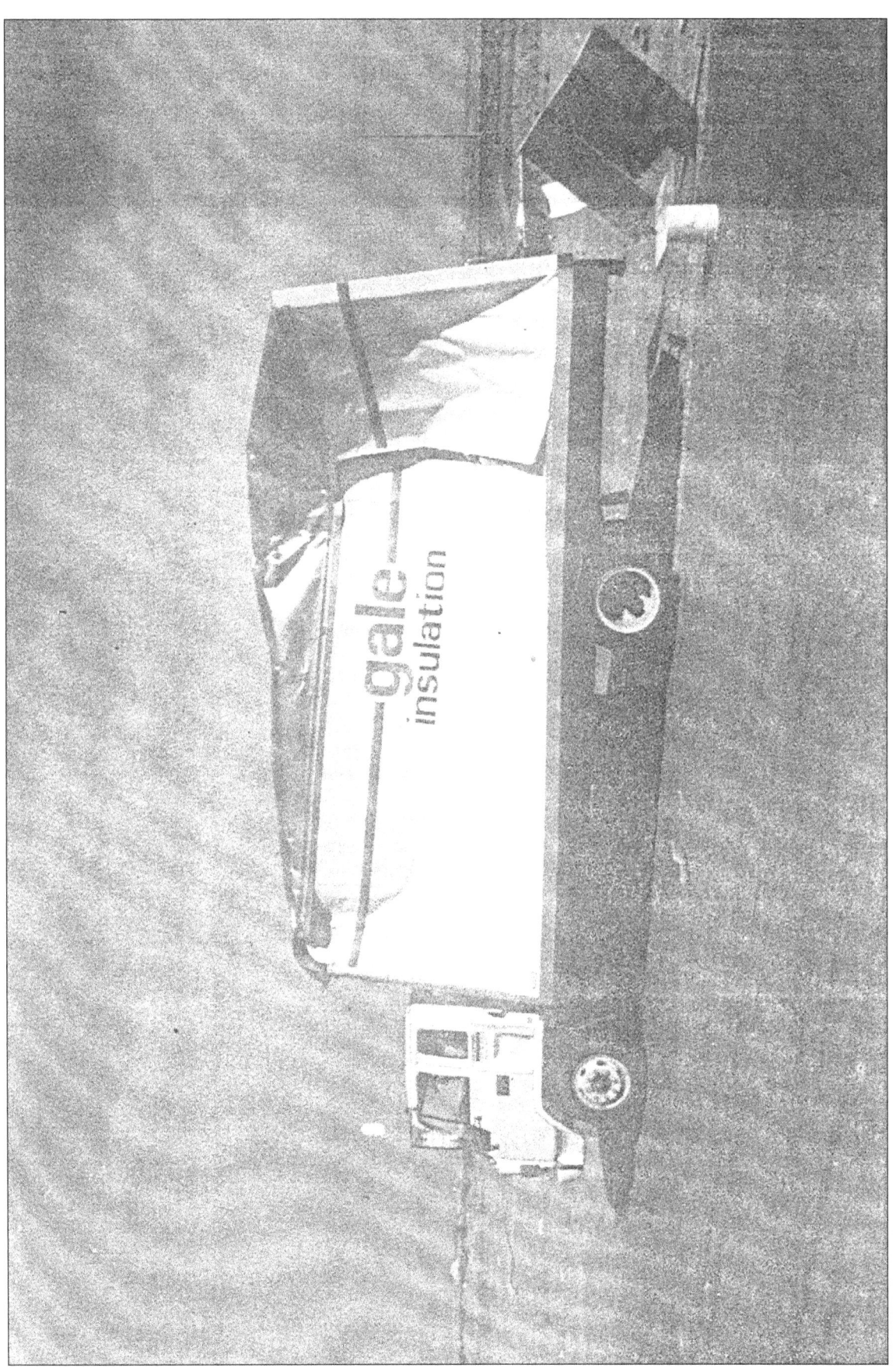

20. Truck damaged by blasts (same truck as in Photo 6).

21. Railroad crossing – approximately 1/2 mile from PEPCON Plant.

22. Damage to warehouse roof – approximately 1 mile from PEPCON Plant.

23. Damage to warehouse roof – approximately 1 mile from PEPCON Plant.

www.ingramcontent.com/pod-product-compliance
Lightning Source LLC
Chambersburg PA
CBHW081241170526
45165CB00009B/3144